INVOLVING DISSOLVING

Teacher's Guide

Grades 1–3
(with modifications for K)

Skills
Observing, Comparing, Describing,
Measuring, Recording, Predicting,
Drawing Conclusions

Concepts
Dissolving Solutions, Evaporation, Crystals

Themes
Systems & Interactions, Patterns of Change, Structure, Matter

Nature of Science and Mathematics
Interdisciplinary, Cooperative Efforts, Creativity & Constraints,
Theory-Based and Testable, Changing Nature of Facts and Theorie
Real-Life Applications

Time
Four 15- to 50-minute sessions
plus follow-up activities

Leigh Agler

LHS GEMS

Great Explorations in Math and Science (GEMS)
Lawrence Hall of Science
University of California at Berkeley

Illustrations
Lisa Klofkorn

Photographs
Richard Hoyt

Lawrence Hall of Science, University of California, Berkeley, CA 94720. Chairman: Glenn T. Seaborg; Director: Marian C. Diamond

Publication was made possible by grants from the A.W. Mellon Foundation and the Carnegie Corporation of New York. This support does not imply responsibility for statements or views expressed in publications of the GEMS program. GEMS also gratefully acknowledges the contribution of word processing equipment from Apple Computer, Inc. Under a grant from the National Science Foundation, GEMS Leader's Workshops have been held across the country. For further information on GEMS leadership opportunities, please contact GEMS at the address and phone number below.

International Standard Book Number: 0-912511-50-8

COMMENTS WELCOME

Great Explorations in Math and Science (GEMS) is an ongoing curriculum development project. GEMS guides are revised periodically, to incorporate teacher comments and new approaches. We welcome your criticisms, suggestions, helpful hints, and any anecdotes about your experience presenting GEMS activities. Your suggestions will be reviewed each time a GEMS guide is revised. Please send your comments to: GEMS Revisions, c/o Lawrence Hall of Science, University of California, Berkeley, CA 94720. The phone number is (510) 642-7771.

Great Explorations in Math and Science (GEMS) Program

The Lawrence Hall of Science (LHS) is a public science center on the University of California at Berkeley campus. LHS offers a full program of activities for the public, including workshops and classes, exhibits, films, lectures, and special events. LHS is also a center for teacher education and curriculum research and development.

Over the years, LHS staff have developed a multitude of activities, assembly programs, classes, and interactive exhibits. These programs have proven to be successful at the Hall and should be useful to schools, other science centers, museums, and community groups. A number of these guided-discovery activities have been published under the Great Explorations in Math and Science (GEMS) title, after an extensive refinement process that includes classroom testing of trial versions, modifications to ensure the use of easy-to-obtain materials, and carefully written and edited step-by-step instructions and background information to allow presentation by teachers without special background in mathematics or science.

Staff

Glenn T. Seaborg, Principal Investigator
Robert C. Knott, Administrator
Jacqueline Barber, Director
Cary Sneider, Curriculum Specialist
Rosita Fabian, Kimi Hosoume, Carolyn Willard,
        Staff Development Specialists
Cynthia Ashley, Administrative Coordinator
Gabriela Solomon, Distribution Coordinator
Lisa Haderlie Baker, Art Director
Carol Bevilacqua and Lisa Klofkorn, Designers
Lincoln Bergman and Kay Fairwell, Editors

Contributing Authors

Jacqueline Barber
Katharine Barrett
Lincoln Bergman
David Buller
Fern Burch
Deborah Calhoon
Linda De Lucchi
Jean Echols
Alan Gould
Sue Jagoda
Robert C. Knott
Larry Malone
Gay Nichols
Cary I. Sneider
Elizabeth Stage
Jennifer Meux White

Reviewers

We would like to thank the following educators who reviewed, tested, or coordinated the reviewing of this series GEMS materials in manuscript form. Their critical comments and recommendations contributed significantly to these GE publications. Their participation does not necessarily imply endorsement of the GEMS program.

ARIZONA

David P. Anderson
Royal Palm Junior High School, Phoenix
Joanne Anger
John Jacobs Elementary School, Phoenix
Cheri Balkenbush
Shaw Butte Elementary School, Phoenix
Flo-Ann Barwick Campbell
Mountain Sky Junior High School, Phoenix
Sandra Caldwell
Lakeview Elementary School, Phoenix
Richard Clark*
Washington School District, Phoenix
Kathy Culbertson
Moon Mountain Elementary School, Phoenix
Don Diller
Sunnyslope Elementary School, Phoenix
Barbara G. Elliot
Tumbleweed Elementary School, Phoenix
Joseph M. Farrier
Desert Foothills Junior High School, Phoenix
Mary Anne French
Moon Mountain Elementary School, Phoenix
Leo H. Hamlet
Desert View Elementary School, Phoenix
Elaine Hardt
Sunnyslope Elementary School, Phoenix
Walter Carroll Hart
Desert View Elementary School, Phoenix
Tim Huff
Sunnyslope Elementary School, Phoenix
Stephen H. Kleinz
Desert Foothills Junior High School, Phoenix
Alison Lamborghini
Orangewood Elementary School, Phoenix
Karen Lee
Moon Mountain Elementary School, Phoenix
George Lewis
Sweetwater Elementary School, Phoenix
Tom Lutz
Palo Verde Junior High School, Phoenix
Midori Mits
Sunset Elementary School, Phoenix
Brenda Pierce
Cholla Junior High School, Phoenix
Sue Poe
Palo Verde Junior High School, Phoenix
Robert C. Rose
Sweetwater Elementary School, Phoenix
Liz Sandberg
Desert Foothills Junior High School, Phoenix
Jacque Sniffen
Chaparral Elementary School, Phoenix
Rebecca Staley
John Jacobs Elementary School, Phoenix
Sandra Stanley
Manzanita Elementary School, Phoenix
Chris Starr
Sunset Elementary School, Phoenix
Karen R. Stock
Tumbleweed Elementary School, Phoenix
Charri L. Strong
Mountain Sky Junior High School, Phoenix
Shirley Vojtko
Cholla Junior High School, Phoenix
K. Dollar Wroughton
John Jacobs Elementary School, Phoenix

CALIFORNIA

Carolyn R. Adams
Washington Primary School, Berkeley
Judith Adler*
Walnut Heights Elementary School, Walnut Creek
Gretchen P. Anderson
Buena Vista Elementary School, Walnut Creek
Beverly Braxton
Columbus Intermediate School, Berkeley
Dorothy Brown
Cave Elementary School, Vallejo
Christa Buckingham
Seven Hills Intermediate School, Walnut Creek
Elizabeth Burch
Sleepy Hollow Elementary School, Orinda
Katharine V. Chapple
Walnut Heights Elementary School, Walnut Creek
Linda Clar
Walnut Heights Elementary School, Walnut Creek
Gail E. Clarke
The Dorris-Eaton School, Walnut Creek
Sara J. Danielson
Albany Middle School, Albany
Robin Davis
Albany Middle School, Albany
Margaret Dreyfus
Walnut Heights Elementary School, Walnut Creek
Jose Franco
Columbus Intermediate School, Berkeley
Elaine Gallaher
Sleepy Hollow Elementary School, Orinda
Ann Gilbert
Columbus Intermediate School, Berkeley
Gretchen Gillfillan
Sleepy Hollow Elementary School, Orinda
Brenda S.K. Goo
Cave Elementary School, Vallejo
Beverly Kroske Grunder
Indian Valley Elementary School, Walnut Creek
Kenneth M. Guthrie
Walnut Creek Intermediate School, Walnut Creek
Joan Hedges
Walnut Heights Elementary School, Walnut Creek
Corrine Howard
Washington Elementary School, Berkeley
Janet Kay Howard
Sleepy Hollow Elementary School, Orinda
Gail Isserman
Murwood Elementary School, Walnut Creek
Carol Jensen
Columbus Intermediate School, Berkeley
Dave Johnson
Cave Elementary School, Vallejo
Kathy Jones
Cave Elementary School, Vallejo
Dayle Kerstad*
Cave Elementary School, Vallejo

Diane Knickerbocker
Indian Valley Elementary School, Walnut Creek
Joan P. Kunz
Walnut Heights Elementary School, Walnut Creek
Randy Lam
Los Cerros Intermediate School, Danville
Philip R. Loggins
Sleepy Hollow Elementary School, Orinda
Jack McFarland
Albany Middle School, Albany
Betty Maddox
Walnut Heights Elementary School, Walnut Creek
Chiyomi Masuda
Columbus Intermediate School, Berkeley
Katy Miles
Albany Middle School, Albany
Lin Morehouse*
Sleepy Hollow Elementary Schoool, Orinda
Marv Moss
Sleepy Hollow Elementary School, Orinda
Tina L. Neivelt
Cave Elementary School, Vallejo
Neil Nelson
Cave Elementary School, Vallejo
Joyce Noakes
Valle Verde Elementary School, Walnut Cree
Jill Norris
Sleepy Hollow Elementary School, Orinda
Janet Obata
Albany Middle School, Albany
Patrick Pase
Los Cerros Intermediate School, Danville
Geraldine Piglowski
Cave Elementary School, Vallejo
Susan Power
Albany Middle School, Albany
Louise Rasmussen
Albany Middle School, Albany
Jan Rayder
Columbus Intermediate School, Berkeley
Masha Rosenthal
Sleepy Hollow Elementary School, Orinda
Carol Rutherford
Cave Elementary School, Vallejo
Jim Salak
Cave Elementary School, Vallejo
Constance M. Schulte
Seven Hills Intermediate School, Walnut Creek
Robert Shogren*
Albany Middle School, Albany
Kay L. Sorg*
Albany Middle School, Albany
Marc Tatar
University of California Gifted Program, Berkeley
Mary E. Welte
Sleepy Hollow Elementary School, Orinda
Carol Whitmore-Waldron
Cave Elementary School, Vallejo
Vernola J. Williams
Albany Middle School, Albany
Carolyn Willard*
Columbus Intermediate School, Berkeley

Mary Yonekawa
The Dorris-Eaton School, Walnut Creek

KENTUCKY

Joyce M. Anderson
Carrithers Middle School, Louisville
Susan H. Baker
Museum of History and Science, Louisville
Carol Earle Black
Highland Middle School, Louisville
April B. Bond
Rangeland Elementary School, Louisville
Sue M. Brown
Newburg Middle School, Louisville
Donna Ross Butler
Carrithers Middle School, Louisville
Stacey Cade
Sacred Heart Model School, Louisville
Sister Catherine, O.S.U.
Sacred Heart Model School, Louisville
Judith Kelley Dolt
Gavin H. Cochran Elementary School,
Louisville
Elizabeth Dudley
Carrithers Middle School, Louisville
Jeanne Flowers
Sacred Heart Model School, Louisville
Karen Fowler
Carrithers Middle School, Louisville
Laura Hansen
Sacred Heart Model School, Louisville
Sandy Hill-Binkley
Museum of History and Science, Louisville
Deborah M. Hornback
Museum of History and Science, Louisville
Patricia A. Hulak
Newburg Middle School, Louisville
Rose Isetti
Museum of History and Science, Louisville
Mary Ann M. Kent
Sacred Heart Model School, Louisville
James D. Kramer
Gavin H. Cochran Elementary School,
Louisville
Sheneda Little
Gavin H. Cochran Elementary School,
Louisville
Brenda W. Logan
Newburg Middle School, Louisville
Amy S. Lowen*
Museum of History and Science, Louisville
Mary Louise Marshall
Breckinridge Elementary School, Louisville
Theresa H. Mattei*
Museum of History and Science, Louisville
Judy Reibel
Highland Middle School, Louisville
Pamela R. Record
Highland Middle School, Louisville
Margie Reed
Carrithers Middle School, Louisville
Donna Rice
Carrithers Middle School, Louisville
Ken Rosenbaum
Jefferson County Public Schools, Louisville
Edna Schoenbaechler
Museum of History and Science, Louisville
Karen Schoenbaechler
Museum of History and Science, Louisville
Deborah G. Semenick
Breckinridge Elementary School, Louisville
Dr. William McLean Sudduth*
Museum of History and Science, Louisville
Rhonda H. Swart
Carrithers Middle School, Louisville

Arlene S. Tabor
Gavin H. Cochran Elementary School,
Louisville
Carla M. Taylor
Museum of History and Science, Louisville
Carol A. Trussell
Rangeland Elementary School, Louisville
Janet W. Varon
Newburg Middle School, Louisville

MICHIGAN

Glen Blinn
Harper Creek High School, Battle Creek
Douglas M. Bollone
Kelloggsville Junior High School, Wyoming
Sharon Christensen*
Delton-Kellogg Middle School, Delton
Ruther M. Conner
Parchment Middle School, Kalamazoo
Stirling Fenner
Gull Lake Middle School, Hickory Corners
Dr. Alonzo Hannaford*
Western Michigan University, Kalamazoo
Barbara Hannaford
The Gagie School, Kalamazoo
Duane Hornbeck
St. Joseph Elementary School, Kalamazoo
Mary M. Howard
The Gagie School, Kalamazoo
Diane Hartman Larsen
Plainwell Middle School, Plainwell
Miriam Hughes
Parchment Middle School, Kalamazoo
Dr. Phillip T. Larsen*
Western Michigan University, Kalamazoo
David M. McDill
Harper Creek High School, Battle Creek
Sue J. Molter
Dowagiac Union High School, Dowagiac
Julie Northrop
South Junior High School, Kalamazoo
Judith O'Brien
Dowagiac Union High School, Dowagiac
Rebecca Penney
Harper Creek High School, Battle Creek
Susan C. Popp
Riverside Elementary School, Constantine
Brenda Potts
Riverside Elementary School, Constantine
Karen Prater
St. Joseph Elementary School, Kalamazoo
Joel Schuitema
Woodland Elementary School, Portage
Pete Vunovich
Harper Creek Junior High School, Battle
Creek
Beverly E. Wrubel
Woodland Elementary School, Portage

NEW YORK

Frances P. Bargamian
Trinity Elementary School, New Rochelle
Barbara Carter
Jefferson Elementary School, New Rochelle
Ann C. Faude
Heathcote Elementary School, Scarsdale
Steven T. Frantz
Heathcote Elementary School, Scarsdale
Alice A. Gaskin
Edgewood Elementary School, Scarsdale
Harriet Glick
Ward Elementary School, New Rochelle
Richard Golden*
Barnard School, New Rochelle

Seymour Golden
Albert Leonard Junior High School, New
Rochelle
Don Grant
Isaac E. Young Junior High School, New
Rochelle
Marybeth Greco
Heathcote Elementary School, Scarsdale
Peter C. Haupt
Fox Meadow Elementary School, Scarsdale
Tema Kaufman
Edgewood Elementary School, Scarsdale
Donna MacCrae
Webster Magnet Elementary School, New
Rochelle
Dorothy T. McElroy
Edgewood Elementary School, Scarsdale
Mary Jane Motl
Greenacres Elementary School, Scarsdale
Tom Mullen
Jefferson Elementary School, New Rochelle
Robert Nebens
Ward Elementary School, New Rochelle
Eileen L. Paolicelli
Ward Elementary School, New Rochelle
Donna Pentaleri
Heathcote Elementary School, Scarsdale
Dr. John V. Pozzi*
City School District of New Rochelle, New
Rochelle
John J. Russo
Ward Elementary School, New Rochelle
Bruce H. Seiden
Webster Magnet Elementary School, New
Rochelle
David B. Selleck
Albert Leonard Junior High School, New
Rochelle
Lovelle Stancarone
Trinity Elementary School, New Rochelle
Tina Sudak
Ward Elementary School, New Rochelle
Julia Taibi
Davis Elementary School, New Rochelle
Kathy Vajda
Webster Magnet Elementary School, New
Rochelle
Charles B. Yochim
Davis Elementary School, New Rochelle
Bruce D. Zeller
Isaac E. Young Junior High School, New
Rochelle

DENMARK

Dr. Erik W. Thulstrup
Royal Danish School of Educational Studies,
Copenhagen

*Trial test coordinators

Contents

Date Due

Acknowledgments

These activities were developed by Jacqueline Barber and Kimi Hosoume at the Lawrence Hall of Science as part of a series of early childhood chemistry classes. Over the years, Leigh Agler and other staff of the Chemistry Education Department have presented these activities to hundreds of children and teachers. These presentations, along with the comments of teachers who tested these activities, resulted in the refinements and extensions found in this guide. Special thanks to Jacqueline Barber during the final refinement process for her diligent and tireless crusade against gelatin fungus.

Introduction

Sparkling mystery crystals are added to a hot, colored liquid. Slowly the "island" of crystals seems to melt away. The mixture is given a stir, and the crystals disappear!

The most common phenomena, when looked at with fresh and curious eyes, often turn out to be among the most fascinating. Next time you add sugar to your tea, make lemonade, or add salt to soup, take a closer look and ask yourself, "What is really happening to those crystals?"

The concept of *dissolving* can be grasped at different age levels. Even very young students can observe how a dissolving solid "disappears" when mixed with a liquid, thus supporting the definition of dissolving as "when a solid seems to disappear." However, reaching the understanding that the solid is **still in the liquid,** even though it can't be seen, is more difficult.

In *Involving Dissolving* your students conduct experiments in which substances dissolve and then are "brought back." They make solutions, and watch with inquisitive eyes as liquids evaporate and solids crystallize. They gain experience, discovering that some substances dissolve while others do not; some dissolve quickly, while others take days. In the process, important science skills are practiced and refined, including observing, comparing, describing, recording, and predicting results.

Before beginning the unit, we recommend that you refer to "Helpful Hints for Hands-On Science" on page 50. An extra pair of hands (and ears) to assist you will be helpful in many of these activities. A place in your room where students can meet, away from the materials at their desk, helps facilitate discussion. A sample letter you may copy (or adapt) and send home with students to collect many of the common materials needed for this unit is also provided on page 64.

CONTINUED

We also strongly recommend that you try out each of these activities by yourself before presenting them to your class. In this way, you will be able to anticipate problems that may arise, conduct demonstrations and step-by-step procedures more smoothly, and adjust volume measurements and other variables to suit your own classroom conditions. A "Behind the Scenes" section contains some scientific background that may be of interest to you. It is not intended to be read out loud to your students. The "Resources" section includes other teaching materials that may prove helpful. "Summary Outlines" are also provided to assist you in guiding your students through these activities.

So, start your students measuring, mixing and stirring up some wonder, as they learn how involving dissolving can be!

Time Frame

Activity 1: Homemade "Gel-o"

Teacher Preparation:	20 minutes
Classroom Activity:	30 minutes

Activity 2: Gelatin Disks

Teacher Preparation:	25 minutes
Classroom Activity:	40 minutes

Activity 3: Starry Night

Teacher Preparation:	35 minutes
Classroom Activity:	50 minutes
(daily observations for 1-2 weeks)	

Activity 4: Disappearing Eggshells

Teacher Preparation:	15 minutes
Class Activity:	three 15-minute sessions
(daily observations for 1-2 weeks)	

Activity 1: Homemade "Gel-o"

Overview

A solid is added to a liquid and "disappears." Where did it go? This first activity invites your students to propose all sorts of imaginative theories to explain this commonplace event. The emphasis is on encouraging students to wonder about something they may have seen many times before, but never really pondered. Your students observe what happens when a "mystery solid" (gelatin powder) and a "mystery liquid" (hot fruit juice) are mixed, and how the result changes over time.

For the purposes of this activity, dissolving is defined as "when a solid seems to disappear in a liquid." If your students have not had much experience describing solids and liquids, you may wish to precede this activity with others that are more focused on developing classification and description skills. See "Resources" on page 49 for some suggestions.

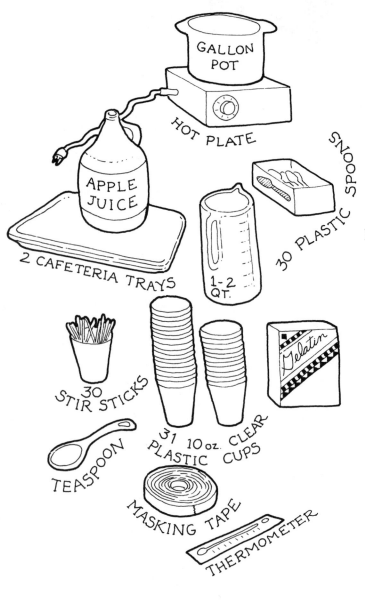

What You Need

For the class:

☐ 1 hot plate, electric coffeemaker, microwave oven or enough thermos bottles to hold several quarts of hot liquid

☐ 1 gallon (4 liter) pot (if you heat the liquid using a hot plate)

☐ 1 or 2 quart-size (1–2 liter) pitchers

☐ 2 cafeteria trays

☐ 1 roll of masking tape

☐ 1 teaspoon

☐ 1 thermometer

☐ 1 clear plastic 10 oz. cup

☐ sponges or paper towels

☐ chalkboard and chalk

☐ about ⅔ cup fruit juice (for demonstration)

For each student:

☐ 1 clear plastic 10 oz. cup

☐ about ⅔ cup fruit juice (See "Getting Ready" #1, below.)

☐ 1 teaspoon of unflavored gelatin (1/2 an envelope)

☐ 1 stir stick (plastic coffee stirrers work well)

☐ 1 plastic spoon (for the next day)

Getting Ready

1. Choose what kind of fruit juice to use. We recommend using apple juice because it is clear, which allows your students to see the gelatin dissolve in it. Apple juice is also relatively healthy and inexpensive. Be advised that the "Gel-o" produced with unsweetened apple juice is not as sweet as commercial gelatin desserts, causing some children to feel disappointed with the taste of their homemade "Gel-o." If this concerns you, you may substitute any other **clear, light-colored** juice, with the exception of pineapple juice, as it prevents the gelatin from setting. One gallon is enough juice for 24 students.

2. Before the day of the activity, have your students help by labeling their cups. Pass out the cups and strips of masking tape. Collect the cups and line them up on trays.

3. Distribute the gelatin powder in the cups. Measure one teaspoon of powder (or half an envelope) into each cup. **Save the empty gelatin container.** If you are using cups that hold less than 10 ounces, you may need to adjust the amount of gelatin and juice accordingly.

4. Twenty minutes before you introduce the activity, heat the juice to approximately 45° C. This is the temperature of very hot, but not scalding, tap water. **Save an empty juice container.** If you do not have and cannot bring a heat source into the classroom, heat the juice earlier and keep it hot in thermos containers. Teachers with access to microwave ovens have used them to heat pitchers of juice quickly.

5. Assemble the stir sticks, one empty cup, and the pitchers near where you will present the activity.

Introducing the Activity and Safety Rules

1. Tell the students that they will each be given a "mystery solid" and a "mystery liquid" to observe carefully and then mix together.

2. Explain that it's okay to look at, feel, and smell the "mystery solid," but also emphasize the following two safety precautions before passing the cups out.

- Remind the class never to taste unknown substances.

- Demonstrate how to smell the powder safely by passing your hand over the cup to your nose.

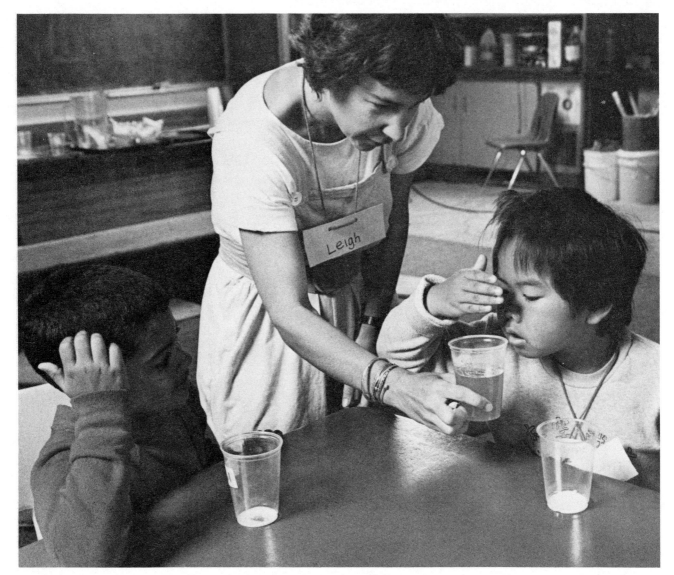

Observing and Mixing the Mystery Substances

1. Tell the class that first they will be given the "mystery solid" to observe. Distribute a cup with the gelatin powder in it to each student.

2. Challenge your students to come up with several different words to describe the solid in their cups. List these observations on the chalkboard under the heading "Mystery Solid." When a comparison is offered as a description, help your students use descriptive words. For example, if they say the solid is like sand, you can ask, "In what ways does it look like sand?" or "What is the same about the mystery solid and sand?"

3. With the class watching, pour some of the hot juice into the extra clear cup. Ask the students to tell you several things they notice about the juice. List these observations on the chalkboard under the heading "Mystery Liquid." Have one student come up and smell the liquid safely, and describe its smell to the rest of the class. Add that description to the list of observations. Save this cup of mystery liquid for the next class session.

4. Remind the students that they will each get some of the same liquid to mix with the solid, and that they should carefully observe what happens. Encourage them to make predictions about what they think might happen. Ask, "Will the liquid change?" "Will the solid change?"

5. Explain that they will each get a stir stick to stir the mystery mixture. If the stir sticks are hollow, you might again want to caution your students not to taste the liquid, as the temptation to suck the liquid through the sticks is great.

6. Fill each student's cup about half full with the hot juice and hand the child a stir stick. Some teachers prefer to have a student helper distribute the stir sticks. As you circulate among the students, listen to the descriptions of what they observe. Encourage them to speculate about what happened to the solid.

7. When the students have completely dissolved the gelatin into the liquid, collect the stir sticks.

Where Did It Go?

1. Direct the attention of the class to the list of observations on the chalkboard. Ask the students to report what happened when they mixed the solid with the liquid. You might want to help focus the discussion by asking, "Did the liquid change?" "Did the solid change?" "How?"

2. Explain to the class that there is a special word to describe what they saw. Write *dissolve* on the chalkboard. Tell them that when a solid is mixed with a liquid and the solid seems to disappear, we could also say that the solid has dissolved. You may want to ask one or two students to use the word in a sentence.

3. Many students may say, "it dissolved," without having a real understanding of what is meant. Help clarify their response by asking questions like, "if the powder has dissolved, where do you think the powder is now?" Have several students respond. Various opinions may be expressed, such as, "it's gone," "it's evaporated," "it melted," or even more surprising statements. Encourage all responses. The question is not an easy one! Most importantly, try to assist your students to use their direct observations in making thoughtful guesses. Do not make comments about whether an answer is "right" or "wrong." It is far more powerful for the students to find out from their own experiments! You may want to take a class poll by having students raise their hands as to whether they think the solid is gone, is still in the liquid, has gone somewhere else, or whether they do not know.

4. Explain that one way of finding out if the solid is still in the liquid is to see if the liquid changes. If it changes, then maybe the solid is still there, causing it to change.

5. Collect the cups. If possible, refrigerate them. If not, cover the cups with sheets of paper and store them in a cool place. Set aside the cup of mystery liquid to use in the next session.

The Next Day

1. Bring out the cups of "Gel-o," the cup of mystery liquid, the empty gelatin package, the empty juice container, a pitcher, and spoons.

2. Before distributing the cups, explain to the class that you saved one cup of "mystery liquid" that did not have the "mystery solid" in it, so they could compare it to their cups.

3. Hold up the cup of "mystery liquid," and slowly tip it, letting it pour out into the pitcher. Then hold up one of your students' cups, and slowly tip it, enough for the class to see that it has "gelled."

4. Ask the students why they think the liquid will not pour out. After several students give their opinions, reveal that the "mystery solid" is still in the liquid. It cannot be seen because it has dissolved in the liquid, but one way to know it's there is to see how it has changed the liquid and made it so it does not pour. You may want to add that not all solids change liquids in this way when they dissolve, but this particular "mystery solid" does.

5. By now, your class will probably have guessed that they have made a dessert like what they know as "jello." Reveal the identity of the "mystery liquid" by showing them the empty juice container. Show them the empty box of unflavored gelatin (the "mystery solid") and tell them where it can be purchased, so they can explain to their parents how to perform the experiment at home.

6. Distribute the cups and spoons and have the students sample the desserts they created! You may want to let your students know that their homemade "Gel-o" will taste different than other gelatin desserts they have had, in order to help adjust their expectations. Ask if they can taste the fruit juice in it.

Going Further

1. Try dissolving other solids in water, in rubbing alcohol, or in oil. Try some that will dissolve and others that will not. Try powders (like dry tempera paint or baking powder) and crystals (like sugar or epsom salts), as well as other solids (like paper and metal).

2. Have your students begin a "Dissolving Recipe Book," as a record of all the dissolving experiments they do in this unit. Depending on the experience and ability level of your class, this could be a class project, or students could write their own reports. Or you could have your students draw what they did, what they saw, and what happened.

3. Have your students make up stories in which the main character is a "Mystery Solid" who gets dissolved in a "Mystery Liquid." Have them tell about what happens to the solid as it gets dissolved and what it does to the liquid. Then have them illustrate the story.

4. Experiment by dissolving the same amount of gelatin in different amounts of liquid. How are the results different? Some teachers have made "gelatin blocks" with their students and had them compare the blocks with the homemade "Gel-o." Recipes for gelatin blocks can be found on packages of unflavored gelatin.

Modifications for Kindergarten

1. Before beginning this activity, provide several experiences that help your students classify substances as either solids or liquids. Several of the materials listed in the "Resources" section on page 49 include activities that are excellent for this purpose.

2. As appropriate, you may choose not to record students' observations of the "Mystery Solid" and "Mystery Liquid," or you may want to reduce the number of observations recorded on the chalkboard.

Activity 2: Gelatin Disks

Overview

In this activity, the students are asked to predict what might happen when a little hot water and a lot of gelatin powder are mixed. Using small circular containers, each student mixes the gelatin in warm water, observing as the powder dissolves and the mixture gels. Several drops of food coloring are added.

Within several days or a week, the mixtures harden and can be used to make interesting "stained-glass-like" hanging ornaments.

What You Need

For the class:

☐ 1 hot plate, electric coffeemaker, microwave oven, or a thermos bottle to keep one quart of water hot

☐ 1 pot or tea kettle, with at least 1 quart capacity (if you heat the water using a hot plate)

☐ 1 quart of water

☐ 1 thermometer

☐ 1 small pouring container, such as a measuring cup with a spout or a small pitcher

☐ 2–3 squeeze bottles of food coloring. Different colors are preferable.

☐ a roll of masking tape

☐ newspaper

☐ a one-hole paper punch or glue

For each student:

☐ 1 small plastic, flexible lid with relatively high sides (See "Getting Ready" #1 below.)

☐ 1 envelope of unflavored gelatin (2 teaspoons)

☐ 1 sturdy paper plate (or 2 thinner ones)

☐ 1 stir stick (plastic coffee stirrers work well)

☐ a pencil

☐ 1 piece of yarn, (6" long if the students want hanging ornaments—15" long if the students want necklaces)

Getting Ready

1. Acquire one plastic, flexible lid with relatively high sides for each student. Liver container lids work best and can often be purchased inexpensively from (or donated by) butchers. Lids from cottage cheese or family-sized yogurt containers are often this same size. The instructions in this activity assume that you will use these large lids, which are 4.5 inches in diameter and contain about 50 ml (about 1.5 ounces) of liquid. Don't use lids with paper labels on them.

Other lids, such as those from small yogurt containers, can be used as long as each is of the same diameter. If you have collected smaller lids, you will want to reduce the amount of gelatin and liquid used to make each disk. For instance, if you have a lid that holds only about half this amount of liquid, then you will want to use only half the amount of gelatin. A typical eight-ounce yogurt container lid is "half-size," i.e. it holds only 20–25 ml (about .75 ounces) of liquid. If you alter the recipe, **try it first by yourself** before using it with your students. If not all of the gelatin dissolves, then you should use less gelatin when you do it with your students. If the resulting disk is too thin or tears, then you need to add more gelatin.

2. **It is important that you make a gelatin disk for yourself before having your students make them.** In humid climates and weather conditions, the disks will take longer to dry and will occasionally grow mold. This will not occur in most situations. There are some measures you can take to reduce the mold growth if you know, after trying it yourself, that mold growth will be a problem. See "Behind the Scenes" on page 45 for a discussion of what can be done.

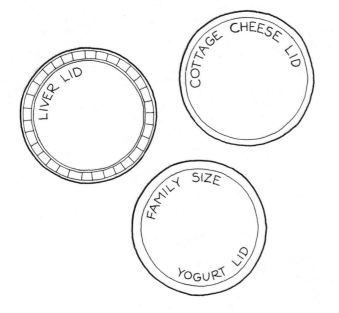

3. Stick a strip of masking tape to the underside of each lid. Later, students will write their names on this tape.

4. Clear off a flat shelf or cabinet top on which you can place and store all the lids for one week or longer.

5. Right before class begins, heat the quart of water to approximately 45° C, so it is hot, but not scalding. Very hot tap water is often hot enough. Add two tablespoons of vinegar to the water and stir. This will help prevent the growth of mold while the gelatin disks are hardening.

6. Have students cover their desks with newspaper before the activity.

Introducing the Activity

1. Review what happened during the "Homemade Gel-o" activity by asking:

- In the last activity, what did you see when you added the gelatin powder to the hot juice?

- What happened the next day?

- How could you tell that the gelatin powder was still in the juice, even though you couldn't see it?

- Do you remember the word we used to describe what happened when the gelatin powder disappeared into the liquid? (Write the word *dissolve* on the chalkboard.)

2. Remind the class that in the last activity, they mixed a small amount of gelatin powder into half a cup of liquid. Write the word *solution* on the chalkboard. Tell them that when a solid dissolves in a liquid, the mixture is called a **solution.**

3. Explain that today they will be making another solution, but this time they will use a lot of gelatin powder and only a little bit of water. Challenge your students to predict what might happen. Have three or four students make predictions.

You may wish to encourage the entire class to participate by using some of the predictions in a straw poll, such as: "Raise your hand if you think a solution made from a lot of gelatin and a little water will get harder than it got last time. Raise your hand if you think it will be clear. Raise your hand if you think something else will happen. Raise your hand if you don't have a guess."

Teacher Demonstration

1. If possible, gather your students in a place away from their desks so they can focus on your instructions.

2. Place a lid on a paper plate. Explain to your students that they will be using the lid as a dish, and that the paper plate is to contain any spills. Show them how to place the lid so its small walls make it like a dish. Write your name on the paper plate and tell them they will do the same.

3. Also demonstrate writing your name on the strip of masking tape on the underside of the lid. (Alternatively, an indelible marker can be used to write directly on the inside of the lid.)

4. Show the students how to empty all of the gelatin powder from the envelope into their lid.

5. Explain that you will then come around to each of them and add some hot water to fill their lid. Do this now and show them how to stir the mixture until all the gelatin disappears. Emphasize that they should stir gently so as not to spill their solutions or get hot water on themselves. **Remind your students not to taste the solution.**

6. Tell the students that after their gelatin powder is all or mostly dissolved, you will come around with food coloring to make their solutions look more interesting. Put one drop of each of two or three different colors of food coloring in your solution. Explain that if you were to stir the coloring or add too much, it would turn brown and look muddy. Show them how to gently turn their paper plates around without lifting the plates off the table, so that the coloring swirls slowly in the solution.

PAPER PLATE

GELATIN DISK

DON'T LIFT THE LID

Mixing

1. As students return to their desks, distribute the lids, paper plates, and pencils and have them begin writing their names on their plates and lids.

2. Distribute an envelope of gelatin powder and a stir stick to each student. As you circulate around the room, check to make sure that all students have their lids "dish-side-up" in the middle of the paper plate, that their names are written on the underside of their lids, and on their paper plates.

3. Have the students empty the gelatin powder into their lids.

4. Using the small pitcher, go around to all students, filling their lids with hot water. Remind them to stir the solution gently, until the gelatin powder is dissolved.

5. Now go around to all the lids, adding two or three drops of food coloring. **Collect the stir sticks as you go,** or have a student go ahead of you to do so. Remind the students to turn their paper plates carefully to swirl the colors in their solutions.

Reflecting

1. Again, ask, "What happened to the gelatin powder?" You may want to encourage discussion by polling the class in this way: "Raise your hand if you think the gelatin powder is still in the water." "Raise your hand if you think it is somewhere else."

2. Ask the students, "How might we find out if the gelatin is still in the water?" If the idea doesn't surface on its own, reinforce the concept from the last activity that one way to tell if the gelatin is still there is to look for changes in the water. Suggest to the class that the lids be left for a few days to see if any changes take place.

3. Check a few of the disks. If they have gelled enough so they will not spill if carried, have the students carry their disks on the paper plates to the shelf or counter you've cleared off.

In the Days that Follow

1. The solutions in the disks may take anywhere from three days to two weeks to harden completely, depending on the size of the lids, weather conditions, and other factors. When it hardens, the gelatin disks will look clear and can be separated from the lids.

2. Show the class a few examples of the hardened disks. Ask your students: "How has this changed?" "Do you think the gelatin powder was in the water when we left these on the shelf?" "Do you think the powder is still there now?" "How can you tell?"

3. Pass out the lids to the students. Usually, they will be able to lift out the disks by carefully peeling back the lid. Some disks may be harder to remove, and you may need to assist by making a small cut into the lid with scissors or a knife to help bend the lid back.

4. The disks will have irregular, thin edges. If your students use scissors, have them trim the edge to make it smooth. If they do not use scissors, you may want to do this for them.

5. Use a paper punch to make holes in the disks so your students can tie yarn loops through the holes to hang the disks. Or, you can use glue to attach the yarn. The disks can then be hung in a classroom window, used for necklaces, or taken home for decorations.

Going Further

1. Have your students write a short report on "Gelatin Disks" to add to their "Dissolving Recipe Notebook."

2. Make a class mobile to hang from the ceiling using the gelatin disks.

Modifications for Kindergarten

1. Write student names on the lids and plates yourself.

2. Have materials set up on students' desks. The time immediately following recess or lunch is good because you can distribute materials while students are away from their desks. Plan to introduce the activity in an area away from student desks. Let the students know where they should meet, away from their desks, when they return.

3. Do not introduce the word "solution" as part of the scientific vocabulary. Instead, use the word "mixture."

Activity 3: Starry Night

Overview

In this activity, the students measure, mix, and observe one substance that dissolves in water and another that does not. Then they filter their solutions and pour them onto squares of black paper. In the days following the experiment, the wet papers begin to dry, and students notice that crystals begin to form. In the weeks that follow, students observe more crystals emerging on the black paper, creating the impression of a beautiful "starry night."

If your students are ready for other new concepts, this activity has wonderful teaching opportunities for introducing and reinforcing the concepts of *evaporation* and *crystal.*

2 CAFETERIA TRAYS

4 1 LITER CLEAR PLASTIC POP BOTTLES

4 ONE GALLON PLASTIC JUGS

SCISSORS

WATER PITCHER

16 COTTAGE CHEESE CONTAINERS

CRYSTAL KOSHER SALT

10" COFFEE FILTERS

30 STIR STICKS

STYROFOAM MEAT TRAY

35 CLEAR PLASTIC CUPS

COARSE GROUND PEPPER

MAGNIFYING LENSES

NEWSPAPERS

2 LARGE SHEETS BLACK CONSTRUCTION PAPER

ABOUT 16 WHITE OR YELLOW CRAYONS

What You Need

For the class:

- ☐ 4 one-gallon plastic jugs, such as milk jugs
- ☐ 4 clear, colorless plastic soda bottles, 1 liter volume
- ☐ 20 coffee filters, about 10 inches in diameter, flat or cone-shaped (paper towels may be substituted)
- ☐ 1 or 2 water pitchers, 1 quart capacity or larger
- ☐ 1 styrofoam meat tray
- ☐ 2 cafeteria trays or 15 styrofoam meat trays
- ☐ 2 sheets of 8-1/2" x 11" black construction paper
- ☐ 1 cup kosher, pickling, or sea salt (these salts are preferable because when dissolved they leave no visible trace. The additives in most other varieties of table salt cause water to look cloudy. However, if additive-free salt is unavailable, any table salt will work.)
- ☐ ½ cup coarsely-ground pepper (finely-ground pepper is unsuitable)
- ☐ 5 clear plastic cups, 6-10 oz.
- ☐ chalkboard and chalk or a large sheet of butcher paper
- ☐ newspapers
- ☐ paper towels or sponges

For each team of four students:

- ☐ 2 small containers, such as cottage cheese containers, or ice cream dishes
- ☐ 2 teaspoons (or teaspoon-sized plastic spoons)

For each student:

- ☐ 1 stir stick
- ☐ 1 clear plastic cup, 6-10 oz.
- ☐ 1 light-colored crayon, such as white or yellow

Optional:

- ☐ a magnifying lens

Getting Ready

1. Make the filtering funnels and containers in the following way:

   a. Cut the tops from the gallon jugs at the point where the walls begin to curve (see illustration). These will serve as funnels.

   b. Cut the bottoms from the plastic soda bottles at the point where the walls begin to curve. Soak off the labels. These will serve as large, clear, containers.

2. Cut small pieces of black construction paper, about 2 inches by 2½ inches, one for each student. Also make one extra piece for demonstration purposes.

3. Fill half of the small containers with 15 teaspoons of salt each. Fill the remaining containers with 6 teaspoons of pepper.

4. Choose a large table or shelf of table height on which the funnels may be used. Cover the shelf or table surface with newspaper. Set aside another area where the trays can be left undisturbed for one to three weeks.

5. Fill the pitchers with water. In the area where you plan to introduce the activity, place the water pitchers, five clear cups, a stir stick, teaspoons, and the containers of salt and pepper. If there is not a chalkboard in this area, post the sheet of butcher paper.

6. Rearrange classroom desks or tables and chairs so groups of four students can share materials. Cover the desks with newspaper.

Introducing the Activity

1. If possible, gather your students in a place away from their desks so they can focus on your instructions.

2. Remind the class that in the last two activities they observed gelatin powder dissolve in different amounts of liquid. Ask the class, "Do you think all solids will dissolve when they're put in a liquid?" Have several students respond.

3. Tell your students that in today's activity they will be doing an experiment to find out what happens when two other solids are mixed with a liquid.

4. Explain that the two solids are very familiar ones, salt and pepper. Ask your students to describe how salt looks, feels, and smells. Write their descriptions of salt on the chalkboard. Then put a pinch of salt into two of the clear cups and pass these to two different students. Ask them to verify or modify the descriptions the class made, and add these observations to the board.

5. Repeat the same process for pepper, asking the class to describe it, recording their descriptions on the board, passing out some pepper to two students, and then writing their observations on the chalkboard. If you have magnifying lenses available, you may want to have several students examine the salt and pepper with the lenses to elaborate upon the descriptions.

6. Help the class summarize their observations by asking, "How are salt and pepper different?" How are they the same?"

7. Demonstrate the following procedure:

    a. Use a spoon and stir stick to measure three level teaspoonfuls of salt into a clear cup.

    b. Measure one level teaspoonful of pepper, and put it in the same cup with the salt.

    c. Use the stir stick to gently mix the salt and pepper together.

8. Have the group repeat to you the amounts of salt and pepper that you mixed in the cup. Tell them that the salt and pepper containers will be shared by four students, and have them return to their desks.

Experimenting

1. Distribute the salt and pepper containers, spoons, stir sticks, and plastic cups.

2. Have each student put three spoonfuls of salt and one of pepper into his cup and gently mix.

3. Then walk around the room with the pitcher and add water to each of the student's cups. Ask students to indicate for you the halfway level on their cups and fill each cup to that halfway point. Remind them to stir gently. As you circulate, ask the students whether or not the two solids, salt and pepper, are dissolving in the water.

4. When you see that most students have dissolved the salt in their cups, ask several students to collect the stir sticks and the salt and pepper containers. Ask the class for their results: "Raise your hand if your pepper dissolved." "Raise your hand if your salt dissolved." "Raise your hand if both solids dissolved." Usually students will agree that the salt dissolved (mostly) and the pepper did not (mostly).

5. Involve the students in a discussion about what happened. Ask the students, "Where did the salt go?" "How might we take the undissolved pepper out of the water?"

Funneling and Filtering

1. Tell your students that they will be using a funnel and a filter to remove the pepper from their solutions. Show the class the homemade funnels, filters, and containers. Explain that when they pour their solutions in the paper-lined funnel, the water will pass through into the container and the pepper will stay in the funnel. This is a way scientists sometimes separate solid pieces from a mixture of solid and liquid.

2. Demonstrate how they should carry their cups very carefully, with two hands, to the funnels set up in the classroom, and slowly pour the solutions into the container. If you have assistants, have them help hold the funnels when your students pour. If you do not have helpers, you may want to help several students pour and then have them hold the funnels for the others.

3. Before beginning the filtering, give the students something to do while they are waiting their turns. Distribute the black squares of paper and light-colored crayons, such as white or yellow, and have them write their names on both sides of the pieces of black construction paper. They could also start working on short reports for their "Dissolving Recipe Notebook." Students can write or draw about what they did at the top of a page and later glue their "Starry Night" to the bottom of the page.

4. Have groups of students, four at a time, come up to the filters and empty their cups.

5. When all the students have poured their solutions into the containers, refocus the attention of the entire class by holding up one of the filter papers. Circulate around the room as needed so everyone gets a chance to see the filter paper, and ask, "What do you see on the filter paper?"

6. As students conclude that the pepper is on the paper, ask questions like, "Where do you think the salt went?" "Can anyone see the salt in the water?" "Does anyone have a suggestion for ways we could find out whether or not the salt is still in the water, or if it is somewhere else?" Take a number of student suggestions for experiments. If no one suggests it, remind them that in the last two activities they let their solutions sit for several days to see if anything changed.

Preparing the Black Paper

1. Explain that in the next experiment their solutions will be poured onto the black paper on which they wrote their names and they will watch for changes over time. Ask your students to predict what might happen.

2. Collect the pieces of black paper and lay them out on the trays. Have your students watch as you pour the solution over the papers, just enough to completely cover them.

3. Tell the class that you will also make an additional tray with a piece of black paper, but you will pour plain water on it, not the solutions that they made. Ask a student volunteer to write "water" on the black paper and pour plain water over it. Explain that this might be one good way to test to see if there is a difference between plain water and the solutions they made. You may want to mention that scientists often design experiments like this.

4. Have several students carefully help move the trays to a place where they can be left undisturbed for one to three weeks.

5. At this point, you have a choice. If the black paper soaked in the salt solution is left to itself, over time the salt will indeed form some crystals on the paper. However, while some students may have a fair amount of crystals formed, others may have just a few and be quite disappointed. The problem is that due to the process involved in crystallization, the results are much more striking if you "seed" the crystal formation by providing surfaces on which the crystals can form. This is usually done by sprinkling a few salt crystals over the papers when your students are away from the room. This "seeding" will ensure the success of crystal formation and make the "starry night" metaphor more apt. However, some teachers may prefer not to seed in this fashion; others may want to do so but explain what they did to the class after the results are evident. In any case, the results are sufficient to indicate that the salt was dissolved in their solutions and reappeared as crystals when the water evaporated.

Discussing Results

1. In the days that follow, have students check on the progress of the black paper trays. Designate three or four students each day to be responsible for giving the class a report on anything they observe happening. You may want to have them write these descriptions and post them near the experiment. Remind them to compare the papers soaked in the solution with the one soaked in plain water.

2. As students report that the papers are drying, you could ask, "Can you think of another time when you saw something wet become dry?" (Some common experiences with evaporation are: rain-soaked pants drying, bath towels drying overnight, disappearing rain puddles, the water level going down in a fish tank, hair drying.) Ask the students where they think the water has gone.

You may want to take this opportunity to further discuss the concept of *evaporation.* Like the concept of dissolving, evaporation has two different levels at which it can be understood. That the water "disappears" is not hard to understand as it is directly observable by your students. That the water goes into the air as invisible particles of water, or water vapor, is not observable, and therefore difficult for your students to believe. Some teachers find that just concentrating on dissolving is challenging enough for their students. Others choose to capitalize on this teaching opportunity. If you have the time, the experiment described in item #1 of the "Going Further" activities on page 34 suggests a way to introduce evaporation in a more direct way.

3. The black papers will take from one to three weeks to dry completely, depending on the temperature and humidity of the room, and the amount of solution in the trays. As your students begin to report the formation of crystals on their black papers, have them pay particular attention to describing them in as much detail as possible. Ask questions to help focus their descriptions, such as: "Tell me about what you see on your papers." "What color are these things?" "Can you see through them?" "What shapes do you see?" "Are they round-sided or sharp-sided?" "Do they all have the same shape or is each of them different?" Explain that these shapes forming on their papers are called *crystals.* Ask your students to speculate on where these crystals came from and what they might be.

If you want to go more deeply into this subject, you could use their descriptions and the class discussion to help the class arrive at a definition of crystals as, "solids with flat sides and regular shapes." Salt crystals tend to be cubic in shape. Crystals of other substances are often different shapes. Some common crystals are: sugar, salt, rock salt, epsom salts, snow, and quartz. Some non-crystals are: baby powder, dry clay, flour, cornstarch, and dry tempera paint. Item #2 of the "Going Further" activities on page 34 suggests an activity to enrich your students' understanding of crystals.

4. When the papers are completely dry, return them to the students. The crystals on black paper can be compared to stars in a clear night sky. Students can also color yellow "moons" in the corner of the paper, or you could pass out yellow dot stickers for the same purpose, to help complete the image of a "starry night."

Going Further

1. During the time that the black papers are drying, place a few clear cups upside down over some of the papers. Ask your student observer teams to describe anything they notice about the cups. Observations of water drops forming inside the cups will lead to questions about how the drops got there. This will help provide evidence for the statement that water goes into the air when it evaporates. For further evidence, try placing two cups over two plates of water. Put one of the plates in a warm, sunny spot, and the other in a closet. When water drops appear inside the cup in the sun, compare it to the cup in the closet.

2. To assist in further defining crystals, you could have the class compare crystals with powders. On trays or newspaper, line up some examples of crystals on one side of the room, and examples of non-crystals on the other side. Have the students examine both the crystals and non-crystals. If available, hand out magnifying lenses.
After students return to their desks, have them tell you the differences they observed and record their findings on the chalkboard. If there are disagreements, appoint an "impartial" committee of three students to re-examine both groups and reach a conclusion.

3. Try dissolving other solids in water and create a list of solids that dissolve in water and those that do not.

Modifications for Kindergarten

1. Before beginning this activity, provide several unguided experiences with dissolving and nondissolving.

2. Arrange to have an aide, parent, grandparent, or older students available on the day of the activity to help collect cups, hold funnels, and assist in other ways.

3. Have materials set up on students' desks. The time immediately following recess or lunch is good because you can distribute materials while students are away from their desks. Plan to introduce the activity in an area away from student desks. Let the students know where they should meet, away from their desks, when they return.

4. Double the number of salt and pepper containers so that pairs of students can share materials, rather than groups of four.

5. While students are waiting for their solutions to be filtered, you may want to have them draw a picture of what they saw in their cups when they mixed the salt, pepper, and water.

6. Emphasize the idea that some substances dissolve in water, while others do not. Focus the attention of the students on the activity itself, rather than on the subjects of evaporation and crystals. Depending on the level of your class, you may even want to limit the activity to salt alone, rather than introducing the non-dissolving variable of pepper.

Activity 4: Disappearing Eggshells

Overview

Will an eggshell dissolve? This activity, which begins as a demonstration, challenges the students to use the concept and process skills developed in previous activities to understand a different dissolving situation — one that takes place more slowly.

Eggs, both raw and hard-boiled, are placed in vinegar and water. The results are observed over a one- to two-week period. The students soon notice that the eggs immersed in vinegar begin to lose their shells. Later, the solution is poured off and allowed to evaporate. Lo and behold — the calcium from the eggshells slowly recrystallizes in the days that follow.

What You Need

- [ ] 4 eggs
- [ ] 1.5 cups of white vinegar
- [ ] 1.5 cups of water in a pouring container
- [ ] 4 clear plastic cups, 6–10 oz.
- [ ] 2 small trays, such as styrofoam meat trays (or 2 clear pyrex baking dishes)
- [ ] 1 sheet of black construction paper, 8½" × 11"
- [ ] 6 sheets of white paper
- [ ] 3 crayons or markers of different colors
- [ ] 1 plastic bag
- [ ] 1 pair of scissors

WATER

2 STYROFOAM MEAT TRAYS

6 SHEETS of WHITE PAPER

3 CRAYONS

4 CLEAR PLASTIC CUPS

WHITE DISTILLED VINEGAR

4 EGGS

SCISSORS

1 SHEET BLACK CONSTRUCTION PAPER

PLASTIC BAG

Getting Ready

1. Hard boil two of the eggs. Use a pencil to mark the raw eggs with a large "X."

2. Bring the cups, the eggs, the water, the vinegar, white paper, and the crayons to the demonstration area.

3. Cut the black paper into two large letter shapes, one shaped like "W" for water, and one shaped like "V" for vinegar. These letters should be as big as possible but still be able to fit on the trays.

Setting Up the Experiment

1. Gather the class in an area where everyone will be able to see the eggs clearly. You may want to use two sheets of white paper as a backdrop for the cups with eggs in them, to aid visibility.

2. Ask the class to recall the solids they have dissolved in class experiments so far. Hold up an egg and ask the question, "Do you think an eggshell will also dissolve in a liquid?" Have several students respond.

3. Tell the class that they are going to do four different experiments with eggs, and they will need to observe what happens carefully over the next few weeks.

4. Explain that you are now going to begin the four experiments and that you want them to watch you carefully, because after you finish you are going to ask them how the experiments differ from each other.

5. Put the four cups on the demonstration table. Slowly pour water into two of the cups and vinegar into the other two. If the containers do not make it clear, tell the students what the liquids are. Then put the four eggs in, with hard-boiled eggs in both water and vinegar, and the same for the uncooked eggs. Make sure the "X's" on the raw eggs are visible to the students.

6. Ask the class to tell you what is in each of the four cups. When students point out the "X," explain that eggs that are uncooked are marked with an "X," while eggs that are hard-boiled are unmarked.

7. Have four students in the class make signs to tape next to each of the cups: "WATER + COOKED EGG," "VINEGAR + COOKED EGG," "WATER + UNCOOKED EGG," "VINEGAR + UNCOOKED EGG."

8. Students will probably be eager to make guesses about whether or not they think the cooked or uncooked eggshells will dissolve and in which liquids. You may want to conduct a class opinion poll by pointing to each cup and asking students to raise their hands if they think the eggshell in that experiment will dissolve. Be sure to give students the option of not making a guess. You could count the number of hands raised for each one, and write that number on the signs.

9. Place the four cups with their identifying signs on a shelf where the students are able to see them clearly, but also where they will not be disturbed. You may want to mention to the students that moving or otherwise interfering with the cups could affect the results, so they should only observe.

What Happened?

1. Assign "observer teams" to report any changes that take place inside the cups over the next week. You may want to have them write or draw what they see and post the observations near the cups. If the weather is warm, you might want to store the cups in the refrigerator, and bring them out each day for the students to examine. Alternatively, you can cover the containers, so any odor that does develop will be contained during most of the experiment. If you want to avoid all odors completely, then do this experiment with the cooked eggs only.

2. In several days, the shells of the eggs left in vinegar will have dissolved, while the shells of the eggs in water will be unaffected. If the raw eggs have not been disturbed, the entire egg should be visible and intact, held together by the thin outside membrane. By carefully rubbing the white residue (a small amount of remaining calcium) from the outside of the membrane, you and your students should be able to see the raw yolk suspended in the raw egg white.

3. Bring all of the eggs, the black paper letters, two small trays, and a plastic bag to a location where all students will be able to see, and gather your students there.

4. Briefly review the observation reports of the student teams. Have the students summarize the results, emphasizing which eggs still have shells and which do not. Your students will be anxious to investigate the eggs "without shells." At first appraisal, many students will conclude that the cooked egg in vinegar "turned to rubber." Cutting the egg in half or comparing it with the cooked egg in water (with its shell removed) will help them understand that the egg is hard-boiled, and that the cooked egg in vinegar is like the cooked egg in water in every way except for it is missing its shell.

Occasionally the membrane surrounding the egg will confuse the students. You may need to remove this membrane, and show them that it is the same membrane that can be seen under the shell of the cooked egg in water. When you and your students examine the uncooked egg in vinegar, this membrane will be quite evident. Some of the vinegar passes through the membrane into the egg, causing the egg to look bloated and the membrane to be stretched taut.

5. Ask your students, "where do you think the shells that disappeared have gone?" Have several students respond. Explain that eggshells are made mostly of calcium, and that vinegar can dissolve calcium, but water cannot. That is why the eggshells that were in the vinegar dissolved. The liquid now is a solution of calcium and vinegar.

6. Ask the students, "what do you think will happen if I pour the solution over black paper and let it sit for several days?" Show them the black paper letters and explain that in order to make this a more complete experiment, you are going to pour out the liquid from both the vinegar cups and the water cups and compare results.

7. Carefully drain off some of the liquid from all four cups onto appropriate black paper letters, in the trays or pyrex baking dishes. Place the eggs in the plastic bag and discard them. Leave the trays of solution in a sunny spot to help the liquid evaporate more quickly.

The Return of the Eggshell

1. After the liquids have evaporated, gather your students in the group discussion area.

2. Hold up the paper shaped like a "W" and have several students share their observations of the paper. Hold up the paper shaped like a "V" and do the same. "Where did the crystals on the "V" paper come from?" If your students do not suggest this, tell them that these are calcium crystals from the dissolved eggshell.

3. Help the class summarize the experiments and results by asking: "Tell me what it means when we say something dissolved." "How could we tell that the solids were still in the liquids?"

4. You may want to hold a brief discussion to help the class draw some general conclusions about dissolving, evaporation, and crystals. Ask students to recall all the experiments they've done in this unit and explain what they've learned. You could end by taking student suggestions for other experiments that might reveal more about dissolving.

Modifications for Kindergarten

1. Make the signs to label the cups yourself, before the demonstration.

2. You may wish to reduce the number of experiments to two (cooked eggs in water and vinegar) or one (cooked egg in vinegar).

3. Try dissolving chalk, seashells, or calcite in vinegar. Like eggshells, these substances contain calcium carbonate which dissolves in an acid. See "Behind the Scenes" on page 45.

Going Further

1. Put a bone in vinegar for several weeks. The calcium from the bone will slowly dissolve, leaving rubbery cartilage behind. If the bone is long, you might even be able to tie a knot in it.

2. Try immersing eggs in different liquids, such as lemon juice, cola, or orange juice. Any liquid that is an acid will have some effect on the egg shell, however only something as strong as vinegar will completely dissolve the shell.

Behind the Scenes

Dissolving

Dissolving is a process in which a solid, when mixed with a liquid, seems to disappear. The solid is pulled apart by the liquid into small pieces, too tiny to see. The dissolving solid is referred to as the *solute.* The liquid into which the solid dissolves is called the *solvent.* The mixture of the liquid and the dissolved solid is called a *solution.* Solids that won't dissolve in a particular liquid are referred to as *insoluble* in that liquid. Gases can also dissolve in liquids, as in the case of soda pop, where carbon dioxide is dissolved in flavored water.

In *Activity 1: Homemade "Gel-o,"* a small amount of unflavored gelatin is dissolved in warm fruit juice. In *Activity 2: Gelatin Disks,* unflavored gelatin is dissolved again, but this time twice as much gelatin is dissolved in a small amount of warm water. In the first case, the gelatin dissolves fairly quickly. In the second, it takes several minutes for all of the gelatin to dissolve.

Gelatin, unlike salt, sugar, or many other solutes, is a large protein and therefore has special properties. Because of these special properties, the gelatin forms a syrupy liquid when first dissolved, and then forms a gel. This occurs because the long molecules in the gelatin mix to form a large mass of tangled strands that might be compared to a pile of spaghetti. In Activity 2, where very little liquid is used, the gel is very dense, and, in time, most of the water evaporates, leaving behind a hard, clear, disk. In this case, the molecules of gelatin are still linked in a large matrix, but due to the loss of water, a solid with less flexible properties is formed.

The process by which gelatin gels is obviously more complicated than just plain dissolving — too complicated for most students of this age to understand. Your students will understand best what they can see for themselves. The observable properties of gelatin powder and warm liquid, and what they notice when the two combine, provide the basis for their understanding of dissolving. The changed results (either a cup of jiggly gelatin or a clear, hard, disk) help them understand that the gelatin does not really "disappear" when it dissolves, but remains in the liquid in another form.

In *Activity 3: Starry Night,* your students mix salt and pepper in water, and observe most of the salt dissolve while most of the pepper does not. All of the salt *should* dissolve, if your students stir long enough. In practice, however, some students tire of stirring, observe that there are still a few grains of salt visible in their mixtures, and conclude that this small amount will not dissolve. Conversely, while pepper would seem to be insoluble in water, there must be some water-soluble component in pepper, as the solution turns slightly brown. Later, the water is evaporated from the solution, leaving recrystallized salt behind.

In *Activity 4: Disappearing Eggshells,* an eggshell dissolves in vinegar. This demonstration of dissolving is more abstract than the previous experiments in the sense that it happens slowly, over several days. While the students can see that the shell *has* dissolved, they don't actually see it happening.

Eggshells are made of calcium carbonate. Calcium carbonate will dissolve in an acid, such as vinegar. You and your students will notice the formation of small bubbles of gas on the eggshell, after it is put in the vinegar. This is carbon dioxide gas, which is released from the calcium carbonate as it goes into solution.

Crystals

A crystal is a clear, solid, substance with a regular geometric shape. Each type of crystal has its own shape. Salt crystals are cubic. When students look at table salt, however, they will often observe shapes that are more rounded than cubes, or that look oblong. This is because the salt crystals fracture and their corners get worn down over time. However, when a salt crystal first forms, it will always be cubic. The shape of a crystal reflects the shape of a molecule. Just like salt crystals, salt molecules are cubic.

While crystal *shape* remains constant, *size* varies according to the conditions under which a crystal is formed. The longer the amount of time a crystal has to form, the larger the crystal becomes. When a crystal "grows," layer is added to layer on what is referred to as the original crystal unit cell. This is why "seeding" a solution will result in the formation of more and bigger crystals—a single salt crystal will provide a base on which the subsequent layers are built.

Mold Growth

Our environment is filled with bacteria and mold spores—many of which are not harmful. The water from your tap, the air in your classroom, and your students' hands can all "contaminate" the experiments. The gelatin and water in *Activity 2: Gelatin Disks* provide an ideal growth medium for molds. Since the disks are left to sit for a week or more, the mold has an opportunity to grow. However, in most cases, the disks will harden before any major growth has occurred. Once most of the water has evaporated, any mold growth that does take place will be within the gelatin matrix, and barely noticeable. However, in humid climates and during wet weather conditions, the disks dry more slowly, allowing double or even triple the amount of time for the mold to grow before the disks become dry. In these situations, you might well observe the formation of fuzzy mold colonies on the gelatin disks. This fuzzy mold is unsightly and spoils the "stained-glass" effect of the gelatin disks.

There are a number of measures you can take to limit or eliminate the growth of mold. The trick is to do something that will eliminate or slow mold growth, but will not interfere with the experiment or pose a safety concern. If you suspect that mold growth will be a problem for you, or if mold grew on your sample disk, then choose *one* of the following mold growth prevention measures when you present the activities to your students:

- If you can get enough space in a refrigerator, refrigerate the gelatin disks as they dry. The cold will retard the growth of the mold.

- Add a teaspoon of disinfectant (such as Lysol), heavy duty cleanser (such as 409), or mildew stain remover (such as Tilex), to the quart of warm water you use to pour into students' lids. There will be some odor from any of these additives, so you will have to explain what you added and why to your students. As in all experiments, caution your students not to put their disks in their mouths.

- Spray the lids with disinfectant, heavy duty cleanser, or mildew stain remover. Let the lids dry prior to distributing them to your students.

Resources

Classroom Activities

- SCIIS, Science Curriculum Improvement Study, Delta Education, Nashua, NH

Beginnings
Material Objects
Interaction and Systems
Subsystems and Variables

- SAVI/SELPH, Science Activities for Visually Impaired/Science Enrichment for Learners with Physical Handicaps, Lawrence Hall of Science, Berkeley, CA 1980.

Mixtures and Solutions

- GEMS, Great Explorations in Math and Science, Lawrence Hall of Science, Berkeley, CA 1987.

Liquid Explorations, 7 class sessions, Grades K—3. In this series of activities, students explore the properties of liquids. They play a classification game, observe how food coloring moves through various liquids, compare oil and water drops, then create secret salad recipes and an "ocean in a bottle."

Helpful Hints for Hands-On Science in the Classroom

- **If you can, get a helper!** Though assistance is not always necessary, it is always advantageous. If you are doing something for the first time, have many materials to distribute, are concerned about the possibility of spills, or want to be able to better listen and respond to questions, then you will greatly appreciate the extra help. If you can't enlist an aide, or a parent, grandparent, or interested retiree, consider signing on a few reliable fifth or sixth grade students. If you have a regularly scheduled science time, you could set up a revolving team of older helpers so every session is covered. You, your students, and the helpers will all benefit from the experience.

- If possible, hold listening and discussion parts of the activities away from the place where students do their experiments. Students have a difficult time listening if there are materials distracting them. Find a good place to gather your students for these times. A rug area or a reading corner are good possibilities, or you can gather the students around your desk or around a large table. If none of these areas is available, plan to set up materials on trays so you can more easily collect the materials before the discussion. By the same token, once students are involved in an activity, it is usually better not to give new group directions or ask questions of the entire class. Instead, circulate among the students, interacting with small groups. Of course, brief whole-class instructions or clarification are sometimes necessary and there be other exceptions, but in general the hands-on materials are so "involving" they naturally make it difficult for students to focus on discussion at the same time.

- **Reduce distribution of materials during the activity to a minimum.** Plan to do activities that involve distributing many materials at strategic times, for example, the activities might start just after students have been in another classroom, or at lunch or recess. Let students know before they leave your class where you want them to gather when they return. You and or your helpers can then set up the materials at their desks while they are out of the room. (Another approach is to arrange all of the materials for a team of six to eight students on individual trays and place them on a table or shelf. Then have a representative from each team distribute and collect these trays.

- Begin a collection of multi-purpose, waterproof materials. These can include items such as milk carton bottoms, unbreakable containers with lids, plastic spoons, plastic coffee stir sticks, popsicle sticks, clear plastic soda bottles, gallon plastic jugs, and styrofoam egg cartons for use in your science and math activities throughout the year.

- **Collect the materials you need for the unit early!** At least one month before beginning the unit, look over the materials list and decide which you can get from the school and which you already have. Many of the materials can be collected by the students from their families. Feel free to copy the letter to parents that appears on page 63, and add to it and other materials you need.

Assessment Suggestions

Selected Student Outcomes

1. Students articulate that some solids dissolve, while others don't; and some solids dissolve quickly, while other solids dissolve over days.

2. Students explain that though a dissolving solid *seems* to disappear, it actually remains in the liquid.

3. Students improve their ability to observe and describe substances.

4. Students demonstrate increased confidence in predicting/guessing what might happen in an unknown situation.

5. Students improve their ability to describe changes that take place after solids and liquids are mixed.

Built-In Assessment Activities

Regularly Making Predictions: In all the activities, students make predictions about what they think will happen when a solid is dissolved in a liquid and where they think the solid has gone. The teacher can observe students' willingness to make predictions/guesses, and whether they become more confident over the course of the unit. (Outcome 4)

What Happened?: In each of the four sessions, students have the opportunity to comment on their final "product," such as the Gel-O, Gelatin disks, or Starry Night papers. The students explain what happened in the experiment or describe the changes that took place. The teacher can note how the students articulate their responses and how the responses improve in clarity and detail over the four sessions. (Outcome 5)

Additional Assessment Ideas

Dissolving Other Substances: In the Going Further activity described at the end of Activity 1: Homemade Gel-O, students dissolve other substances such as powders, crystals, and various solids. The teacher can observe how well students draw or write about what they did, what they saw, and what happened. (Outcomes 1, 3)

A Story About a Mystery Solid: In another Going Further activity for Homemade Gel-O, students create stories about a mystery solid that is dissolved in a mystery liquid. They tell what happens to the solid as it dissolves and what dissolving it does to the liquid. They then illustrate the story. These stories may reveal students' concepts about where a solid goes after it seems to disappear, or a student's understanding that some solids dissolve while others do not. Some descriptions can demonstrate richness and detail. (Outcomes 1, 2, 3)

Mystery Solution: Give your students a mystery substance such as Kool-Aid powder or Jello powder. Have them think of words to describe the mystery substance, using as many senses as they can. Ask them to predict what might happen when the substance is mixed with cold water, and then let them test the predictions. (Outcomes 3, 4, 5)

Write a Letter: Students can write or dictate a letter to a younger student explaining what happens to salt when it dissolves in water. (Outcome 2)

Summary Outlines

Activity 1: Homemade "Gel-o"

Getting Ready

1. Choose clear, light-colored fruit juice.
2. Have students label cups.
3. Distribute one teaspoon gelatin powder into each cup. Save empty gelatin containers.
4. Heat juice to 45° C. Save empty juice container
5. Place stir sticks, empty cup, and pitcher near presentation area.

Introducing the Activity and Safety Rules

1. Tell students each will be given a "mystery solid," and "mystery liquid," to observe then mix.
2. Explain that they can look at, feel, and smell the "mystery solid," but:
    a. Never taste unknown substances.
    b. Demonstrate safe smelling method.

Observing and Mixing the Mystery Substances

1. Distribute cups with gelatin in them.
2. Ask students for several descriptions. List on chalkboard.
3. Pour hot juice into extra clear cup. Ask for observations and list on chalkboard. Have one student smell it. Save for next session.
4. Encourage students to predict outcome when the solid and liquid are mixed.
5. Explain that they will get a stir stick for mixing. If necessary, caution against using sticks as straw.
6. Fill all student cups halfway with with warm juice and distribute stir sticks. Circulate, encouraging descriptions.
7. After gelatin dissolves, collect stir sticks.

Where Did It Go?

1. Discuss what happened with class and introduce the word **dissolve** as what happens when a solid is mixed with a liquid and seems to disappear.
2. Ask students where they think the powder has gone. Encourage all responses.
3. Explain that one way to find out where the solid has gone is to see if the liquid changes.
4. Collect cups. Refrigerate if possible or store covered in cool place. Set aside cup of mystery liquid.

The Next Day

1. Bring out cups of "Gel-o," mystery liquid cup, empty gelatin package, the empty juice container, a pitcher, and spoons.
2. Hold up mystery liquid and slowly tip it, pouring it into the pitcher. Then hold up one of the student cups, tipping enough for them to see it has "gelled."
3. Ask students why the gelled liquid will not pour out. Explain that the solid is still in the liquid; it has dissolved and caused a change.
4. Reveal identities of mystery solid and liquid.
5. Hand out cups and spoons and let students sample their "Gel-o."

Activity 2: Gelatin Disks

Getting Ready

1. Acquire plastic lids.
2. Make gelatin disk for yourself first.
3. Stick masking tape strip to each lid underside.
4. Clear off shelf or cabinet top for storage.
5. Heat quart of water to 45° C.
6. Have students cover their desks with newspaper.

Introducing the Activity

1. Review last activity and reinforce understanding of the word **dissolve.**
2. Explain that when a solid dissolves in a liquid the mixture is called a **solution.**
3. Explain today's activity: a lot of gelatin powder and a little liquid. Ask for predictions.

Teacher Demonstration

1. Show students lid and paper plate. Demonstrate writing name on plate and on masking tape.
2. Show students how to empty all the gelatin powder into lid.
3. Explain that you will pour hot water into all lids, demonstrate, then stir gently until it dissolves. Remind them not to taste the solutions.
4. Tell students you will add food coloring to solutions. Show them how to turn the paper plates after drops of coloring are added, so it swirls.

Mixing

1. Distribute lids, paper plates, pencils and have students write their names on plates and lids.
2. Distribute gelatin envelopes and stir sticks.
3. Have students empty powder into lids.
4. Go around, filling lids with hot water. Remind students to stir gently.
5. Add drops of food coloring and collect stir sticks.

Reflecting

1. Ask students what happened to gelatin powder.
2. Ask for ideas on how to find out if gelatin is still in water. Suggest lid be left for a few days.
3. If mixture has gelled enough, have students carry lids to storage area.

In the Days that Follow

1. Show class examples of hardened disks. Discuss changes that take place.
2. Pass out disks to students and have them peel off. Edges can be trimmed.
3. Use paper punch for holes to tie yarn for hanging ornaments.

Activity 3: Starry Night

Getting Ready

1. Make funnels and containers with gallon jugs and plastic soda bottles.
2. Cut small pieces of black construction paper for each student. One extra for demonstration.
3. Fill half of small containers with 15 teaspoons of salt; the remainder with 6 teaspoons pepper.
4. Choose table or shelf and cover with newspaper where funnels will be used. Set aside storage area.
5. Fill pitchers with water and assemble materials.
6. Arrange desks so groups of four can share materials. Cover desks with newspaper.

Introducing the Activity

1. Ask class if all solids will dissolve in a liquid.
2. Explain that they will try to dissolve two familiar solids, salt and pepper.
3. Ask students to describe how salt and pepper look, feel, and smell.
4. Demonstrate this procedure:
   a. Use spoon and stir stick to measure 3 level teaspoons of salt into clear cup.
   b. Measure 1 level teaspoon of pepper and put in same cup.
   c. Use stir stick to gently mix salt and pepper.
5. Have class repeat amounts of salt and pepper. Salt and pepper containers shared by four students.

Experimenting

1. Distribute salt and pepper containers, spoons, stir sticks, plastic cups. Have each student put 3 spoonfuls of salt and 1 of pepper into cup and mix.
2. Add water half full to student cups. Remind to stir gently. Ask if two solids are dissolving.
3. Have students raise hands to see if salt and pepper dissolved. Ask "Where did the salt go?" and "How could we take undissolved pepper out of the water?"

Funneling and Filtering

1. Show students homemade funnels, filters, and containers. Explain how to filter out pepper.
2. Show how to carry cup carefully to funnels.
3. As students wait turns, have them write names on both sides of black construction paper.
4. Have groups of 4 empty cups into funnels.
5. When all groups finish, hold up one of the filter papers and ask class what they see.
6. Ask, "where did the salt go?" Discuss how to find out if salt is still in water. If not mentioned remind students that in previous activities they let solutions sit for several days.

Preparing the Black Paper

1. Explain that solutions will be poured on black paper. Ask students to predict outcome.
2. Lay black paper on trays and have students watch as you pour solution over paper, just covering it.
3. Tell class you made another tray and will pour plain water over black paper.
4. Have students help move trays to place where they can remain for 1-3 weeks.
5. Decide whether or not to "seed" the paper with more salt to increase crystal formation.

Discussing Results

1. Over next days, have students check on trays and report findings.
2. As drying is reported you may want to discuss **evaporation.**
3. As students begin to report crystal formation have them describe in detail. Explain that the shapes forming are called **crystals.** You may want to discuss other crystals.
4. When papers are completely dry, return them to students. Crystals can be compared to stars in clear night sky. Students may want to add yellow moon.

Activity 4: Disappearing Eggshells

Getting Ready

1. Hard boil two eggs. Mark uncooked eggs with "X"s.
2. Bring 4 cups, eggs, water, vinegar, white paper, and crayons to demonstration area.
3. Cut black paper into a V-shape and a W-shape.

Setting Up the Experiment

1. Gather class so all can see eggs clearly. Ask, "do you think an eggshell will dissolve in a liquid?"
2. Explain that you will start four experiments with eggs for them to observe what happens over next weeks.
3. Have them watch carefully, so they will be able to tell you how the experiments differ.
4. Put 4 cups on table. Pour water into two and vinegar into other two. Put hard-boiled eggs into both water and vinegar; the same for the uncooked eggs. Make sure "X"s on uncooked eggs are visible.
5. Have 4 students make signs to label each cup: Water + Cooked Egg, etc.
6. Have students discuss and make predictions about which eggshells will dissolve in which liquids.
7. Place eggs on shelf where they can be seen but not disturbed.

What Happened?

1. Assign "observer teams" to report changes.
2. After several days, bring eggs, black paper letters, 2 small trays, and plastic bag to place all students can see.
3. Have students sum up observer reports and discuss results, emphasizing which eggs have shells and which do not.
4. Ask, "where do you think the shells that disappeared have gone?" Have several students respond.
5. Explain that eggshells are mostly made of calcium, that vinegar dissolves calcium while water does not. Ask, "what do you think will happen if I pour the solution over black paper and let it sit for several days?"
6. Show them black paper letters and explain that you will pour the liquid from both water and vinegar cups on appropriate letters.
7. Drain off some of liquid in cups to the black paper letters. Place eggs in plastic bag to discard. Leave trays in sunny spot so liquid evaporates faster.

The Return of the Eggshell

1. After liquids evaporate, gather students in discussion area.
2. Hold up W-shaped paper and have students share observations, then do same with V-shaped paper. Ask, "Where do the crystals on the V-shaped paper come from?" If not mentioned, tell class these are calcium crystals from the eggshell.
3. Summarize the experiment by discussing dissolving. Ask, "How could we tell the solids were still in the liquids?" Hold general discussion on conclusions from all activities in unit.

Literature Connections

Two of the books listed here depict **crystals dissolving**, one from an ant's perspective and another as the process would appear through a microscope. A picture book featuring a deer licking **salt** provides a nice literary extension to the activities in this unit that involve salt. Another book deals with **evaporation** in the context of the water cycle and water purification. Keep on the lookout for books that help children understand more about the **nature of crystals** and **dissolving, or the use of filtration or crystallization as methods of purification or separation**. Let the GEMS project know about them!

You may also want to refer to the GEMS literature handbook: *Once Upon A GEMS Guide: Connecting Young People's Literature to Great Explorations in Math and Science* for listings of other books under related science themes and mathematics strands, as well as for other GEMS guides, such as *Liquid Explorations*, that deal with concepts connected to those that children explore in *Involving Dissolving*.

Greg's Microscope
by Millicent E. Selsam; illustrated by Arnold Lobel
Harper & Row, New York. 1963
Grades: 2–4

> Greg's father buys him a microscope and he finds an unlimited array of items around the house to observe, even the hair of Mrs. Broom's poodle. The illustrations show the salt and sugar crystals, threads, hair, and other material as it appears to him magnified. Solutions of salt and sugar give him a chance to see crystals dissolve. Although this is not a high tech, state-of-the art representation, the fun and empowering experience of playing with scale are well portrayed.

The Magic School Bus at the Waterworks
by Joanna Cole; illustrated by Bruce Degen
Scholastic, Inc. New York. 1986
Grades: K–6

> When Ms. Frizzle, the strangest teacher in school, takes her class on a field trip to the waterworks, everyone ends up experiencing the water purification system from the inside. Evaporation, the water cycle, and filtration are just a few of the concepts communicated in this whimsical fantasy field trip.

Salt

by Harve Zemach; illustrated by Margot Zemach
Farrar, Straus & Giroux, New York. 1977
Grades: 2–4

> This Russian tale tells of a rich merchant's third son, Ivan the Fool, who discovers an island with a mountain of salt. To market his ship's cargo of salt to a foreign king, he secretly adds salt to the food cooking in the royal kitchen. The story could introduce a discussion of how and why salt enhances the flavor of food. The rest of the story involves a beautiful princess, his evil brothers, and a helpful giant.

Salt Hands

by Jane C. Aragon; illustrated by Ted Rand
E.P. Dutton, New York. 1989
Grades: Preschool–2

> On a moonlit summer night, a young girl awakes to find a deer in her yard. She sprinkles salt in her hands and goes out to stand near it. The deer moves closer, and finally licks the salt from her hand until it is all gone. Lends a nice extension to the activities involving salt by providing an opportunity to discuss the need that animals (and people!) have for salt in their diets.

Two Bad Ants

by Chris Van Allsburg
Houghton Mifflin Co., Boston. 1988
Grades: Preschool–4

> When two curious ants set off in search of beautiful sparkling crystals (sugar), it becomes a dangerous adventure that convinces them to return to the former safety of their ant colony. Illustrations are drawn from an ant's perspective, showing them lugging individual sugar crystals and other views from "the small." Good extension to those activities that deal with sugar, dissolving, and crystals.

Dear Parents,

Your child's class will soon begin a unit on "dissolving." They will be conducting experiments with creative results, while learning and practicing important science skills. The class will investigate which substances dissolve and which do not as they make solutions, let liquids evaporate, and observe solids recrystallize. There are many things that we need for this project, so we are asking for your help. Please see if you have any of the following items, or if you know of others who do. We will need the materials by _____ .

- lids from cottage cheese, family-sized yogurt, or liver containers.

- one gallon plastic jugs, such as milk jugs

- clear, colorless, 2 liter-capacity soda bottles

- old newspapers

- plastic coffee stirrers or popsicle sticks

-

-

-

It will be appreciated if all your donations have been cleaned. Are there people in your neighborhood or where you work who might help? Please pass this note on to others!

Thank you very much for your help,